I0797367

SPACE SYSTEMS
STARS AND THE SOLAR SYSTEM

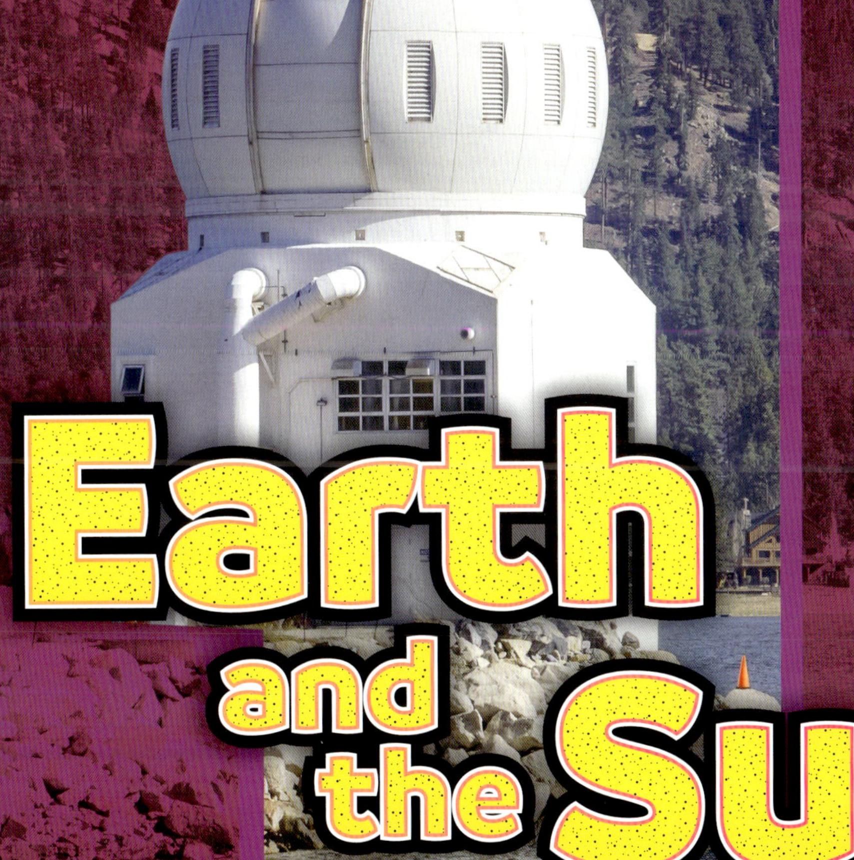

Earth and the Sun

John Willis

LIGHTBOX
openlightbox.com

Go to
www.openlightbox.com
and enter this book's
unique code.

ACCESS CODE

LBXK8787

Lightbox is an all-inclusive digital solution for the teaching and learning of curriculum topics in an original, groundbreaking way. Lightbox is based on National Curriculum Standards.

STANDARD FEATURES OF LIGHTBOX

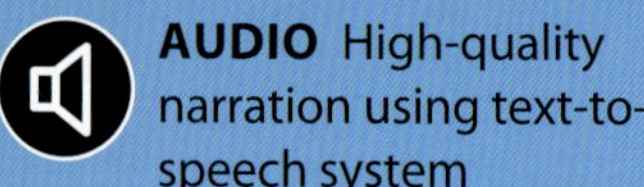

AUDIO High-quality narration using text-to-speech system

ACTIVITIES Printable PDFs that can be emailed and graded

SLIDESHOWS Pictorial overviews of key concepts

VIDEOS Embedded high-definition video clips

WEBLINKS Curated links to external, child-safe resources

TRANSPARENCIES Step-by-step layering of maps, diagrams, charts, and timelines

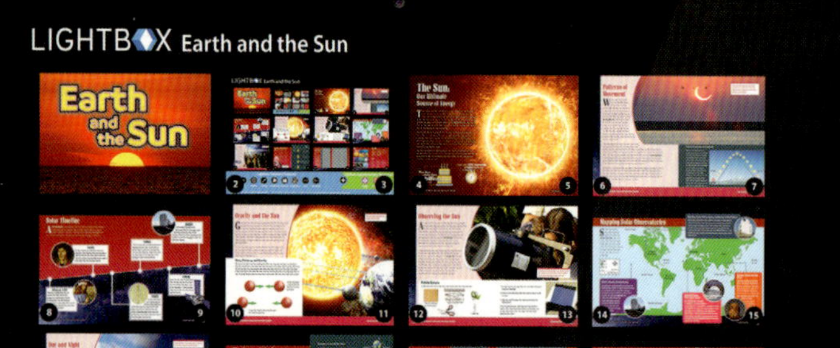

INTERACTIVE MAPS Interactive maps and aerial satellite imagery

QUIZZES Ten multiple choice questions that are automatically graded and emailed for teacher assessment

KEY WORDS Matching key concepts to their definitions

Earth and the Sun

Contents

The Sun:
Our Ultimate Source of Energy

There are countless numbers of stars in the universe. Of these stars, humans are only able to see a tiny number in the sky. The most recognizable star from Earth, and the most important, is the Sun. Like other stars, the Sun is a huge body in space made up of different gases. It is mostly made of the **elements** hydrogen and helium. The Sun formed millions of years ago when **gravity** brought clouds of material in space together.

From Earth, the Sun appears much larger and brighter than other stars. This is not because the Sun is massive. In fact, the Sun is an average-sized star. It appears large because it is much closer to Earth than any other star. Earth is about 93 million miles (150 million kilometers) from the Sun. The next closest star, Proxima Centauri, is 250,000 times farther away.

The Sun constantly radiates energy into space. By doing this, the Sun provides Earth with much of the energy that living things need to survive, in the form of light and heat. Without this energy, there would be no life on Earth.

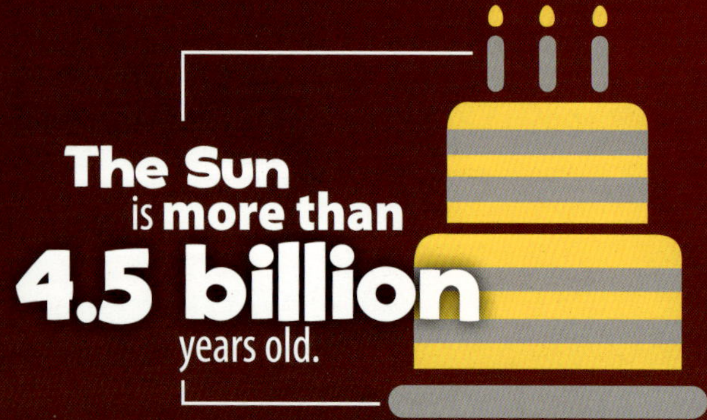

It takes about **500 seconds** for **light** from the Sun to **reach** Earth.

Patterns of Movement

When seen from Earth, the Sun appears quite different from other stars. Unlike other stars, the Sun is visible during the day instead of at night. It often appears as a yellow ball. This is because Earth's **atmosphere** scatters blue light from the Sun, making it appear yellow. During the day, the Sun appears to move across the sky. For many years, people believed that this was because the Sun moved around Earth. Today, people know that this is not the case.

The Sun appears to move across Earth's sky because of Earth's rotation. Earth, like other planets, spins as it travels through space. Each day, Earth rotates from west to east. When this happens, the Sun appears to rise in the east and set in the west. Earth rotates at an angle. At all times, one **hemisphere** is tilted toward the Sun, while the other is not. This means that on parts of the planet tilted away from the Sun, the Sun appears lower in the sky. On other parts of the planet, the Sun seems higher.

When the Sun's light strikes an object, the object leaves a shadow. This is the area where the Sun's light is blocked. When the Sun appears low in the sky, shadows are long. This is because the Sun's light strikes objects from a lower angle. During the middle of the day, the Sun looks high in the sky. This makes shadows much shorter.

When the Moon moves between Earth and the Sun, it blocks some of the Sun's light and casts a shadow on Earth. This is known as a solar eclipse.

From Sunrise to Sunset

What time did the Sun rise today in your city or town? What time did it set? This will change depending on the season and your location. When the Sun rises in the east, it appears red. This is because its light travels through more of Earth's atmosphere before reaching a person's eye, scattering more blue light. Midday, or the solar noon, is the brightest time of day. At this time, the Sun appears highest in the sky. When the Sun sets in the west, it once again appears red.

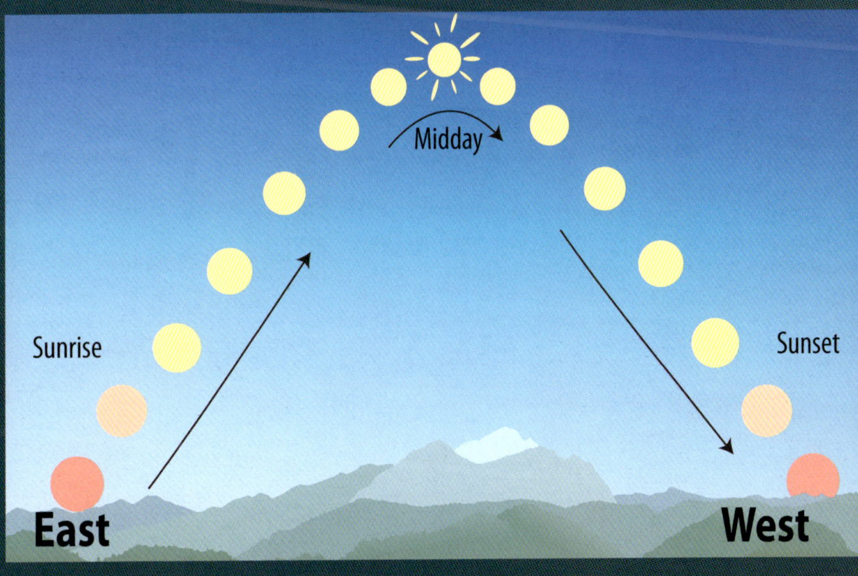

Solar Timeline

Astronomers have studied the Sun since ancient times. Today, they observe Earth's nearest star to learn how it affects people and predict what it may do in the future. Scientists and engineers continue to discover new ways to harness the energy of the Sun.

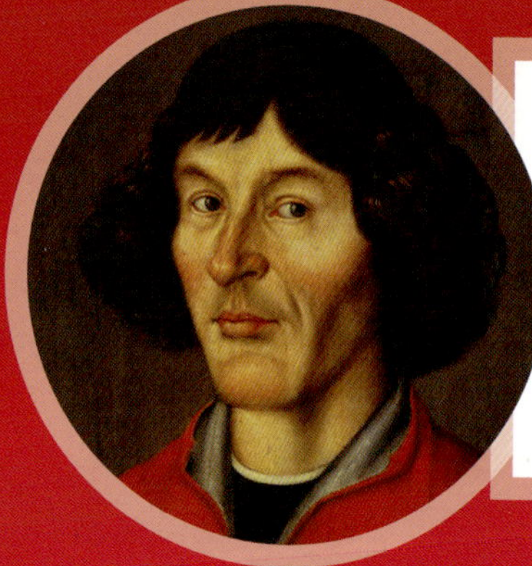

1543

Nicolaus Copernicus, a Polish astronomer, proposes a new model of the solar system in which Earth and the other planets orbit the Sun. This system is much closer to the models used today than the Ptolemaic model.

About 150

Ptolemy, a mathematician and astronomer from Alexandria, proposes a mathematical model of the universe. In the Ptolemaic model, Earth is unmoving at the center of the universe and the Sun, stars, and planets **orbit** around it.

2020

The Daniel K. Inouye Solar Telescope (DKIST), the largest solar telescope on Earth, takes its first pictures of the Sun. These images are the highest-resolution images ever taken of the Sun's surface.

1883

The first solar cell is created by American inventor Charles Fritts. A solar cell is a device that is used to turn energy from the Sun into electricity. Fritts' cell is able to convert about 1 to 2 percent of the energy it receives into electricity.

1958

Solar cells are used to convert the Sun's energy and power systems such as radios on satellites including the Vanguard 1. Today, many spacecraft use solar cells as a power source.

1632

Italian astronomer and mathematician Galileo Galilei publishes a book detailing the arguments for and against Copernicus' model of the solar system. **Controversial** at the time, the publication led to Galileo spending the rest of his life under house arrest.

Gravity and the Sun

Gravity is the force that is responsible for Earth's movement around the Sun. Gravity is a force that any object with mass has. This force pulls objects toward each other. Earth's gravity pulls objects towards its center. This is why, to a person on Earth, the direction of Earth's center is "down." While every object with mass has gravity, it is a very weak force in small objects. In large objects, such as planets, gravity has a stronger effect.

In space, gravity causes smaller objects to orbit larger ones. For example, Earth's gravity is strong enough to keep the Moon orbiting the planet. In turn, the Sun's gravity causes Earth to orbit around it.

Mass, Distance, and Gravity

The mass of an object is not the only thing that affects how its gravity will impact something else. Gravity is also affected by the distance between two objects. For example, Earth is greatly affected by the Sun's gravity, but is not noticeably affected by the gravity of other stars. The larger the distance between two objects, the weaker the force of gravity. Even though other stars may be larger than the Sun, the greater distance between them and the Earth means that Earth will not orbit them instead.

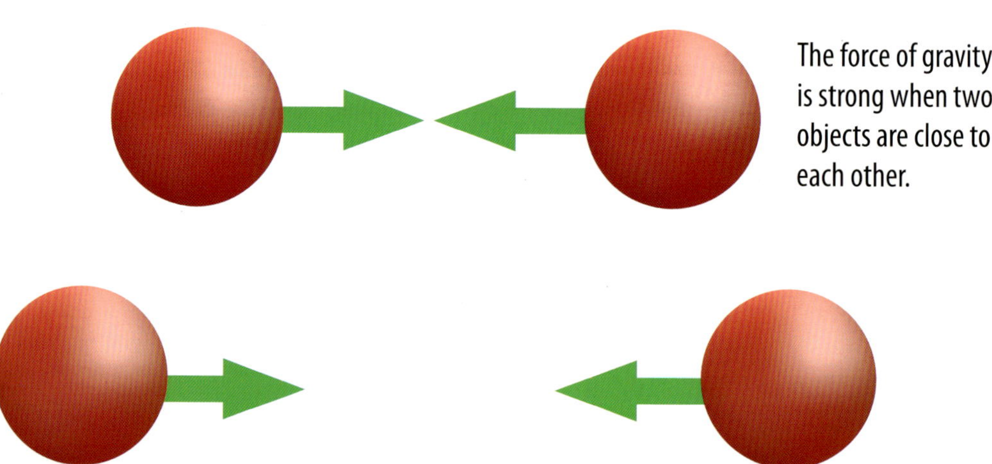

The force of gravity is strong when two objects are close to each other.

The gravity of objects that are farther apart has a much weaker force.

The Sun is about 300,000 times heavier than the Earth.

Observing the Sun

As the Sun is so visible in the sky each day, people have been observing it for all of known history. The Sun is too bright to be observed without any help. When viewed directly, its light can damage the human eye. Staring at the Sun for too long can cause temporary or even long-term **blindness**. This is why it is important never to look directly at the Sun. Due to the danger the Sun presents, people have had to find other ways to observe it.

Today, people use solar telescopes to view the Sun. These are built differently from other telescopes. Most telescopes are designed to **magnify** the light of stars that are very far away. This is not an issue for solar telescopes. Instead, the Sun produces too much light. This is why solar telescopes have filters to block out some of the Sun's light. Another way in which anyone can view the Sun more safely is by using a tool known as a pinhole camera.

Pinhole Camera

A pinhole camera can be made quickly and easily using the following materials:

1 piece of cardboard or cardstock paper

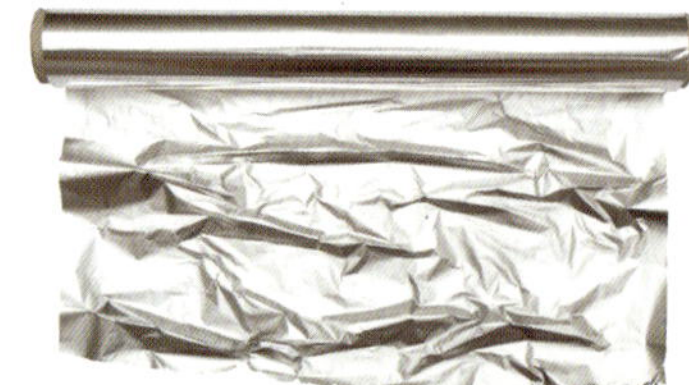

1 piece of aluminum foil

Tape

Scissors

1 paper clip

Solar telescopes allow people to see features, such as sunspots, on the Sun's surface. They also help people view planets and other objects that pass in front of the Sun.

1. Using the scissors, cut a square hole in the middle of the piece of paper or cardboard.
2. Cover the hole using the aluminum foil and use tape to secure it in place.
3. Using one end of the paper clip, make a small hole in the aluminum foil.

To use the pinhole camera, hold it up to the Sun and allow light to shine onto the hole. Place a blank, white object, such as paper, below the camera. The image of the Sun should appear on this surface.

Mapping Solar Observatories

Scientists around the world observe objects in space at **observatories**. Some of these observatories have telescopes or other tools that have allowed scientists to make new discoveries about and observations of the Sun.

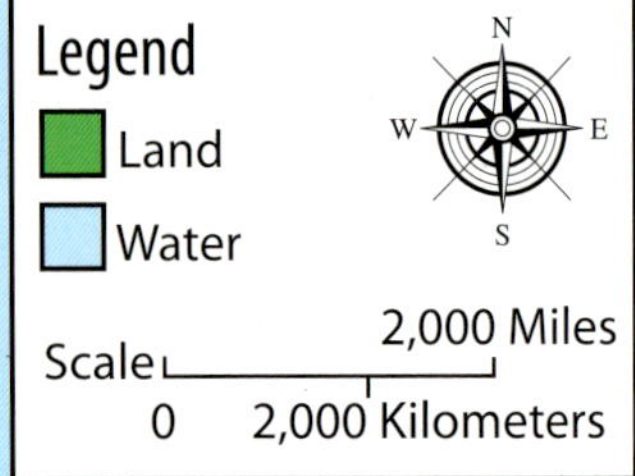

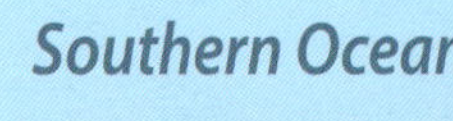

DKIST, Hawai'i, United States

Located on the top of Haleakalā, a volcano on the island of Maui, the DKIST is one of many telescopes at the Haleakalā Observatory. The telescope is located more than 10,000 feet (3,048 meters) above sea level. At such a high **altitude**, less of the Sun's light is blocked by Earth's atmosphere, making it easier to observe. DKIST is designed to view the Sun's magnetic fields and how they affect life on Earth.

Big Bear Solar Observatory, California, United States

The Big Bear Solar Observatory (BBSO) is found on California's Big Bear Lake. The observatory's location helps make its images clearer. This is because the Sun does not heat water as much as it heats land, meaning that there is less moving air above BBSO. BBSO has a range of different tools, including several telescopes, that can be used to view the Sun.

Europe

Asia

Africa

Pacific Ocean

Indian Ocean

Yunnan Observatories, China

Originally established in the late 1930s, the Yunnan Observatories feature several major telescopes, including the New Vacuum Solar Telescope (NVST). The NVST is a vacuum telescope. This means that there is no air at all inside the tube that makes up its body. This allows it to have a clearer view of the Sun than a non-vacuum telescope would.

GREGOR Solar Telescope, Spain

The GREGOR Solar Telescope is located in the Canary Islands, a group of Spanish islands off the west coast of Africa. The GREGOR telescope is the largest European solar telescope. It is an "open" telescope. This means that its body is not surrounded by a tube. Because of this design, air is able to move in and out of the telescope, keeping its **components** cool while it is used to view the Sun.

Day and Night

One of the most important ways in which people experience Earth's orbit around the Sun is the cycle of day and night. As Earth is a sphere, only about one half of the planet faces toward the Sun at a time. This side of the planet is in daylight. The Sun's light shines directly on this side of the planet, causing it to be light and warm.

As Earth rotates, the side of the planet facing the Sun will turn to face away from it. When this happens, that part of Earth experiences night. The Sun does not shine directly on the night side of Earth, so sunlight no longer fills the sky. This makes night darker and cooler than the day. Since the Sun is not visible in the sky, other stars can be seen instead. Their light is not blocked by the brighter light of the daytime Sun.

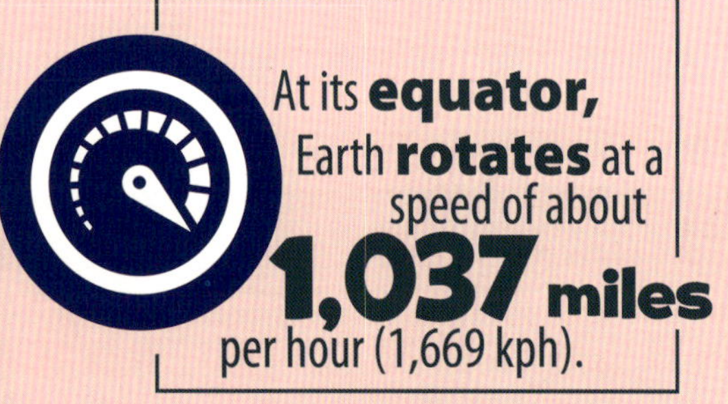

One day on **Venus** is as long as **243 days** on Earth.

It takes 23 hours, 56 minutes, and 4.1 seconds for Earth to complete one full rotation.

The Sun and the Seasons

The average length of each day and night is 12 hours. This is not true all of the time. The length of each day and night changes throughout the year. As these lengths change, areas of Earth experience different seasons.

The reason why the seasons change is because Earth is tilted. This tilt does not change as Earth orbits the Sun. The North Pole is tilted toward the Sun during the months that are summer in the United States. This means that more sunlight shines on the northern half of the planet. Days are longer.

The South Pole is tilted toward the Sun during American winter months. Less light reaches the Northern Hemisphere. This makes days shorter. During fall and spring, neither pole is tilted toward the Sun. These months have days and nights of roughly equal length.

Seasons are different on different parts of Earth's surface. In the Southern Hemisphere, they are the opposite of seasons in the north. At Earth's equator, the effect of the tilt is much less. Day lengths stay nearly the same year-round.

Reduced light in winter does more than just shorten the days. It often causes this time of year to be much colder than summer.

Seasons around the Year

Earth's **axis** is tilted. The planet rotates on this axis as it travels around the Sun. However, the axis always points in the same direction. This is what causes the planet's seasons.

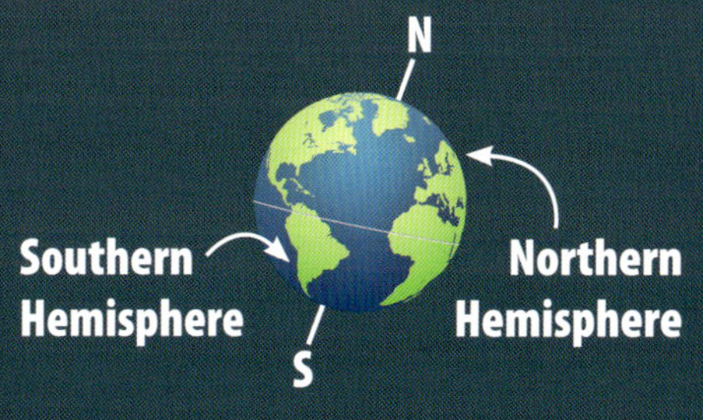

June

The Southern Hemisphere experiences winter, while it is summer in the north. The Sun shines directly on the Northern Hemisphere. Its light indirectly reaches the Southern Hemisphere.

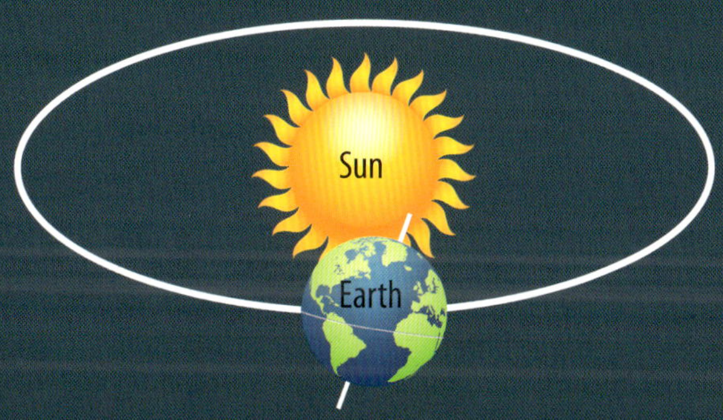

September

The Northern Hemisphere experiences fall, while it is spring in the south. Sunlight reaches both hemispheres equally.

December

The Southern Hemisphere experiences summer, while it is winter in the north. The Sun shines directly on the Southern Hemisphere. Its light indirectly reaches the Northern Hemisphere.

March

The Northern Hemisphere experiences spring, while it is fall in the south. Sunlight reaches both hemispheres equally.

Sunrise and Sunset Chart

The length of days and nights changes throughout the year. You can create a sunrise and sunset chart to track the changes in your area over time.

Materials:
Graph paper
Pen or pencil

1. Using the graph paper, create a chart based on the one here. Label each hour and each month.
2. On the first day of each month, record what time the Sun rises and sets. You can find this out by watching for the sunrise and sunset yourself or by using newspapers or the internet. Alternately, you can choose a year. Using research online, find out when the Sun rose and set on the first day of each of that year's months.
3. Draw a line to connect each sunrise point on the chart with the next one. Do the same for the sunset points. Do the two lines look similar or different? Why do you think this is the case?

	January	February	March	April
11 pm				
10 pm				
9 pm				
8 pm				
7 pm				
6 pm				
5 pm				
4 pm				
3 pm				
2 pm				
1 pm				
12 pm				
11 am				
10 am				
9 am				
8 am				
7 am				
6 am				
5 am				
4 am				
3 am				
2 am				
1 am				
12 am				

May	June	July	August	September	October	November	December

Quiz

1 Which force is responsible for Earth's movement around the Sun?

A: Gravity

2 What causes the Sun to appear to move across Earth's sky?

A: Earth's rotation

3 Is Earth's axis tilted?

A: Yes

4 When was the first solar cell invented?

A: 1883

5 How far is Earth from the Sun?

A: About 93 million miles (150 million km)

6 Is gravity stronger when objects are farther apart?

A: No

7 What do solar telescopes use to block sunlight?

A: Filters

8 In which U.S. state is the Big Bear Solar Observatory located?

A: California

Key Words

altitude: the height of something compared to sea level

astronomers: scientists who study space

atmosphere: a layer of gases surrounding a planet

axis: the point around which an object rotates

blindness: a condition in which a person or animal cannot see

components: individual parts of a larger device

controversial: something that people strongly disagree about

elements: basic substances that cannot be broken down into another kind of substance

gravity: the force that pulls objects with mass together

hemisphere: one side of a planet or moon

magnify: to make something appear larger or more significant

observatories: buildings or areas designed for viewing objects in space

orbit: the path one object in space takes around another

Index

LIGHTBOX

SUPPLEMENTARY RESOURCES

Click on the plus icon ⊕ found in the bottom left corner of each spread to open additional teacher resources.

- Download and print the book's quizzes and activities
- Access curriculum correlations
- Explore additional web applications that enhance the Lightbox experience

LIGHTBOX DIGITAL TITLES

Packed full of integrated media

VIDEOS

INTERACTIVE MAPS

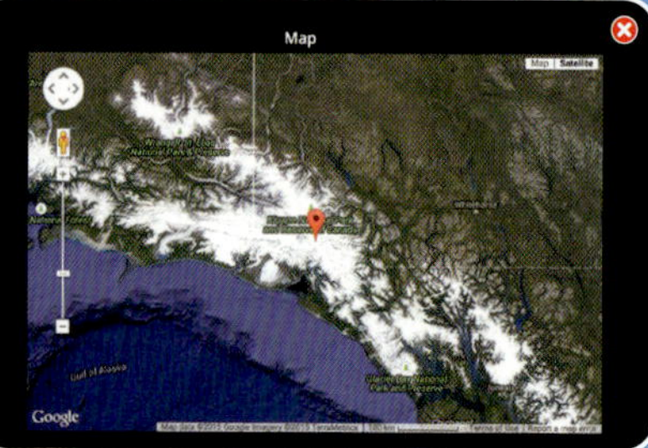

WEBLINKS

SLIDESHOWS

QUIZZES

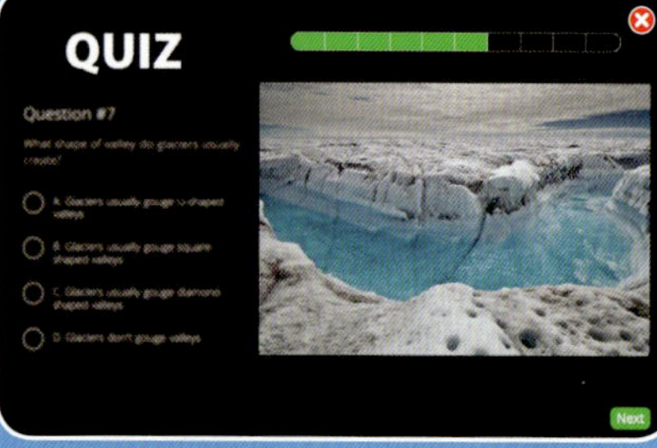

OPTIMIZED FOR

- ✓ TABLETS
- ✓ WHITEBOARDS
- ✓ COMPUTERS
- ✓ AND MUCH MORE!

Published by Smartbook Media Inc.
14 Penn Plaza, 9th Floor New York, NY 10122
Website: www.openlightbox.com

Library of Congress Cataloging-in-Publication Data

Names: Willis, John, 1989- author.
Title: Earth and the sun / John Willis.
Description: New York, NY : Lightbox/Smartbook Media, [2021] | Series: Space systems: stars and the solar system | Includes index. | Audience: Ages 9-14 | Audience: Grades 4-6
Identifiers: LCCN 2020014165 (print) | LCCN 2020014166 (ebook) | ISBN 9781510554726 (library binding) | ISBN 9781510554733
Subjects: LCSH: Sun--Juvenile literature. | Earth (Planet)--Juvenile literature.
Classification: LCC QB521.5 .W557 2021 (print) | LCC QB521.5 (ebook) | DDC 523.7--dc23
LC record available at https://lccn.loc.gov/2020014165
LC ebook record available at https://lccn.loc.gov/2020014166

Printed in Guangzhou, China
1 2 3 4 5 6 7 8 9 0 24 23 22 21 20

062020
111019

Photo Credits
Every reasonable effort has been made to trace ownership and to obtain permission to reprint copyright material. The publisher would be pleased to have any errors or omissions brought to its attention so that they may be corrected in subsequent printings. The publisher acknowledges Getty Images, iStock, and Shutterstock as its primary image suppliers for this title.

Project Coordinator John Willis
Designer Ana María Vidal